Getting Your Bitcoin

A Practical Guide to Obtaining a Bitcoin, Without Buying One

TABLE OF CONTENTS

INTRODUCTION

I'm starting with an introduction to Bitcoin, crypto, and blockchain. If you are already familiar with this, feel free to skip this section or speed read through it. If familiar you can skip right to the, "Blueprint to obtaining a Bitcoin" section.

A cryptocurrency is a digital or virtual currency that is protected by encryption, making counterfeiting and double-spending practically impossible. Many cryptocurrencies are built on blockchain technology, which is a distributed ledger enforced by a distributed network of computers. Cryptocurrencies are distinguished by the fact that they are not issued by any central authority, making them potentially impervious to government intervention or manipulation.

A cryptocurrency is a type of digital asset that is based on a network that spans a huge number of computers. They are able to exist outside of the control of governments and central authorities because of their decentralized structure. Many industries, including finance and law, are expected to be disrupted by blockchain and related technology, according to experts. Cryptocurrencies can either be mined or bought on cryptocurrency exchanges.
Cryptocurrency purchases are of course not available on all eCommerce sites. In fact, even famous cryptocurrencies like Bitcoin are rarely used for retail purchases. Just like you wouldn't take a bar of gold into Starbucks to pay for coffee. Think of bitcoin and most crypto as assets or stores of value.

Blockchain technology is at the heart of Bitcoin's and other cryptocurrencies' attractiveness and usefulness. Blockchain is, as its name implies, a collection of interconnected blocks or an online ledger. Each block comprises a collection of transactions that each network member has independently validated. Every new block must be validated by each node before being confirmed, making forging transaction histories nearly impossible. The contents of an online ledger must be agreed upon by the whole network of a single node, or computer, that keeps a copy of the ledger. According to experts, blockchain technology can benefit a variety of sectors and activities, including supply chain management and online voting and crowdfunding. Large Financial institutions are experimenting with blockchain technology to reduce transaction costs by simplifying payment processing, others are using it to track supply chains and numerous other use cases where keeping exact records is essential.

The most well-known and valued cryptocurrency is Bitcoin. It was conceived and introduced to the public in 2008 by an unidentified individual named Satoshi Nakamoto through a white paper. Thousands of cryptocurrencies are currently available on the market. Every cryptocurrency claims to have a unique purpose and specification. Ethereum's ether, for example, is marketed as gas for the underlying smart contract platform. Banks utilize Ripple's XRP to ease transfers between different locations. The most commonly traded and covered cryptocurrency is Bitcoin, which was first made available to the public in 2009. There were around 18.8 million bitcoins in circulation as of November 2021, with a total market cap of around $1.2 trillion. There will only be 21 million bitcoins ever created. Following Bitcoin's breakthrough, a slew of new cryptocurrencies known as "altcoins" have emerged. Some are Bitcoin clones or forks, while others are brand-new currencies created from the ground up. Solana, Litecoin, Ethereum, Cardano, and EOS are among them.

THE BLUEPRINT TO OBTAINING A BITCOIN

If willing, you are about to embark on a journey with me in earning a bitcoin. First ask yourself, would you rather pay out of pocket and buy one bitcoin or earn it? If you have a boatload of cash hanging around then go buy one it's easier :) If you are like the rest of us who don't have a boatload of extra cash, read on. Now why would you want to come on this journey? How about a million reasons why, starting with many believe that bitcoin could be worth a million dollars or even far more by the 2030s. Not only that but earning a bitcoin is actually fun, it's like a game, a hobby for me because I make it that way. I'd say spending some time earning a bitcoin is definitely worthwhile to say the least. So with that if you're in then say it like you mean it, out loud "I'm in!" and let's get crackin. I'm on a mission to help as many people as possible in this world from every walk of life earn as much bitcoin as possible. So please join me on the journey.

 In this section we will look at understanding the overall idea behind the blueprint to obtaining a Bitcoin. The idea is to do active activities that generate dollars or crypto and then to take dollars or crypto and grow it by using passive activities. The first steps in the blueprint are active activities that generate money that can be converted to passive bitcoin earning. So we are going to first set up and get the active activities going and then you'll keep repeating those activities daily or at least a few times weekly.

Then all that income you generate will be going in on a daily or weekly basis to the passive activities so you can earn bitcoin even while you are on vacation and even when you are sleeping at night or working a day job.

Remember if you don't do something different you'll always get the same results. So if you want to earn a bitcoin, then do as many steps as you can or want in the guide, and you'll be that much closer to living the dream. Keep motivated and consistent by trying to gamify this. We want to actually look at this like something you get up in the morning and you can't wait to see all the ways you can turn things into bitcoin. I literally can't wait for Saturday. I grab a coffee. And I'm paying it forward with this book! Anyway I grab my coffee with my wife and we drive around looking for stuff. Everytime I go Saturday garage saling, all I see is clothing and shoes turning into sats and it's fun! It's a great hobby and it can be addicting. We are all about making the most bitcoin while having the most fun doing it!

So in the next step we are going to cover off on setting up the first passive activity.

The goal is to do whatever it takes to get as much as possible as quickly as possible from Active activities and get a 12% return + on the active side. The reason being is in its good years bitcoin can outpace you, so earn while you can now, and keep it growing along with bitcoin. Take a scenario.

Cathie Wood estimates bitcoin will get a 43% CAGR(Compound Annual Growth Rate) between 2022 and 2032. So with that value. If we can earn around $15,000 and DCA around $500 a month into that scenario monthly from Active activities in 10 years we would likely have a bitcoin or the equivalent of about 1 million dollars.

We can also accelerate this even more than the growth of bitcoin by dividing a portion of those payments and capital across the risk curve to grow a portion of it even quicker than the rates of bitcoin's growth. See the power of getting going on this task right now? If so let's get stackin those sats!

This blueprint is designed like a buffet..if a technique doesn't fit you then simply don't do it and drive onto the next step. You can pick and choose what you want to do but the more you do or like a buffet the more you eat and the more you go for more plates the more full you get...you know what I'm sayin? So complete as many as you can that fit your own research, risk tolerance or style. The only rule is if you are going to do it I highly recommend don't go on to the next step until the step you are on is complete and/or in motion.

Also I realize there are typically two types of learners. Ones that want all the steps and all the details, and for you you'll want to cover off on each step in this blueprint. Then there is the learner that just wants to get quick facts and start putting it into action and get the details later. For you I made a quickstart guide at the very end of the guide. You can print that 1 pager off and hang it on your mirror and start executing on it. To get the most out of this blueprint, or anything out of this blueprint depends 100% on you. You have to be willing to have enough want and desire, to actually take action on these steps. Like motivating yourself to get the sneakers and sweats on and go to the gym, even when you don't feel like it. Game it up and have fun doing this. Be motivated by the future value of bitcoin 5 to years from now. Doing these actions now might be one of the best things you could do. When I don't feel like signing up I ask my future self what I would do for a million dollars and how that would change my life for the better and those that I care for. So let's get ready and get pumped to get that bitcoin.

ONE IMPORTANT NOTE: Programs and offers constantly changed but we have you covered. For updates get the best and latest here. The vehicles like staking, mining, gaming, microtasking won't change or change very little. Google is your friend on getting current and up to date information on the programs going forward.

ACTIVE EARNING ACTIVITIES

Step 1
Action Activity.. Please complete before moving onto **Step 2**

We are starting where we are. The first thing you'll want to do is walk around your house, apartment, yard and look at all the things you already own that you don't really need. Take your phone with you because we are all about taking massive action and doing it immediately so there is no excuse, no procrastination. Open up your phone and download the Mercari application. Mercari is a place that unlike Amazon, Ebay you don't need any fancy listings to get rid of stuff. Just see an item that you would like to convert to Bitcoin :) and snap a photo put in a one liner or two description and you are on your way toward your first step to stacking more of those Bitcoin sats. If you want to, just get a jumpstart and sell a bunch of stuff by holding a garage sale, and put an ad on craigslist, and or your local classified listings. What I did is I told my wife if we haven't seen it or used in 6 months it gets sold. Garage Sale!

We will use all of this active, and semi-active income to constantly fund the passive earning (which will help you grow your bitcoin balance even while you sleep or go to the beach).

Step 2
In this step we are going to reposition paying for things we already do and I'm talking as much as possible into a card that gives us crypto back for our purchases. Get a list together of every bill and if it can be paid by a credit card e.g. groceries, gas, utilities, internet, Netflix/Hulu, phone,

repairs, vacations etc. I don't recommend using services that charge extra so you can pay your mortgage as it sort of defeats the purpose of getting that fee back in our crypto wallets. However, if your lender allows you to pay your mortgage with a credit card then congratulations and all the more crypto back.

Now this is very important. You'll need to make sure that you pay off this card in full typically within 21 to 23 days as you'll incur interest and carry a balance and you don't want that. The idea here is we get money we wouldn't have had coming into crypto that grows over time. We want assets not liabilities.

OK so now that we have IDs on all the bills, now it's time to apply for a crypto back card. In all my research the only legit one I've found that doesn't require you to stake their utility coins, and offers the best cash back is Venmo. Venmo 3% crypto back on biggest purchase, 2% crypto back on 2nd biggest purchase, 1% crypto back on everything else. Now the downside of Venmo is their crypto is in their own wallet, meaning you can't transfer it out to another crypto wallet. They have indicated that they will likely add that, but you are earning real crypto so I still highly like this option. Once you get your Venmo card it's super important that you go into settings and there is an option to set for using your cash back to automatically purchase crypto. It let's you choose bitcoin, ethereum, litecoin or bitcoin cash. I personally have mine set to ethereum and not bitcoin. What! Reason why is ethereum actually grows faster than bitcoin, and so I'm growing my future bitcoin at a much faster rate. I have no problem though, with a bitcoin maxi, just putting it in bitcoin ;) lol.

Before moving along to step 3, please sign up for Venmo and get all your bills set up to automatically pay with Venmo, then set up a reminder or automation to pay that card off in full before the interest accrual date (see your card contract for details).

There is another option for those who have cash on hand to buy crypto and stake it and you can get anywhere from 2% to 8% cash back with crypto dot com. For example, the crypto dot com card requires 4k of CRO coin to get a 3% crypto back reward.

Step 3

Flipping Stuff is the gist of this step. Are you beyond picking up free couches and flipping them? What are you willing to do for that bitcoin that could indeed get to 1 million dollars one day? If you are beyond picking up free stuff and flipping it, then move onto the next technique in step 4. If you are willing to roll up your sleeves and put some sweat equity in for those sats...then here is what we are doing;

Facebook local

Craigslist

Your best local classified ad site

Go and search for keyword: Free

Pickup free stuff then make an ad on Mercari, Facebook local and sell it for an infinite profit.

Another great tactic is just hold a garage sale. Add the free stuff you got to the garage sale and price it to move quickly. Don't get too greedy, give a good deal so you get the money quicker, because that can be getting into bitcoin and growing much faster than the dollar markups over time.

Step 4

Flipping NFTs

This won't be for everyone, but I'll teach you a strategy.

When we were kids we would play a game called bigger or

better. Us kids would get into 2 teams that competed and each team got a single paper clip. We would break up and go knocking on doors and ask people in the neighborhood to give us something bigger or better than the paperclip. Someone gave a watch, who gave a shovel, who gave a bike, who a trampoline etc. At the end of the game the team that one had the most valuable item. Somebody actually took this to an extreme level and ended up with a real house after trading up from a single red paperclip (google it). This same thing applies to NFT flipping. You start small and flip to bigger and better projects. I recommend 2 things with NFTs. The two things are; only deal with what are called Utility NFTs. Utility NFTs give you something of value in the real world, like a meetup, a concert pass, sneakers, a valuable music set etc. The second thing is try to deal with NFTs that are getting a lot of attention. You can find which ones are getting attention right on the blockchain by going to check the transfer volume right on a site called etherscan. You can start free with what are called whitelists, which are early announcements of airdropped NFTs and keep trading up. Coinmarket cap announces NFT airdrops, other sites and discord channels, twitter and telegram are great places to get early info. Eventually we want NFTs with Utility and Attention which is where the big dollars are made. NFT flipping can be extremely risky. The markets can drop 80% off on NFTs with no announcement. That's why unless you have a passion for this as an artist, collector or enthusiast it may not be for you. The strategy you are aiming for is something bigger and better. At some point you'll want to exit in profit and get it into passive profit staking.

Step 5
Almost everyone can make simple videos, write simple articles. I recommend doing something simple called

listicles. You are definitely familiar with what a listicle is because you see them everywhere and the reason you see them is they get a lot of clicks and interest. A listicle is simply a short article, video or content that make a a tops list. Like top 10 places to earn interest on crypto. Top 5 best dog haircut styles, 5 ways to get great abs this summer. Start making some simple listicle videos and articles and monetizing them on places like Youtube, Reddit, publishing content on places like Publish0x, Steemit, Dtube. The great thing with listicles is you can easily modify what's already getting a lot of views for example if top 8 toys for dogs is getting a great deal of hits, make a top 9 toys for cats listicle. Another advantage is these listicles are easy to just publish to multiple platforms and sites.

The more places you post your short video or article too, the more you'll likely earn over time. Many of these sites pay directly in crypto, and some will pay in cash that can easily be converted to crypto on an exchange. Don't move on to step 6 until you try this one! Some of the places to publish are; Publish0x, Getzion, Mediachain, Steem, Rally, and Hive.

Step 6
Paid2 Microtasking. Actively Earn Crypto Directly by doing small tasks called microtasking. This could be anything from watching ads, answering questions, listening to or making music, clicking, walking or exercising and getting paid in crypto to do it. Some of the places to do microtasking are; Smilesbitcoin, Audius, Soldirac, Cointiply, and earncarrot.

Step 7 (let's address the elephant in the room)
Your Job...is valuable. My grandpa used to say the world needs ditch diggers too. He was a master carpenter that did odd side jobs too like picking peaches, janitorial work,

building things and taught us the value of a job, any job is more valuable to you than most give it credit no matter how mennial it may seem. It can actually be your biggest wealth creator.... never discount a job or side hustle..it's how most people became wealthy. If you can put in $100 a check or more and convert it into bitcoin while your bitcoin goes up in value and earns interest while you sleep you can see the power of a job or side hustle. Also in this gig economy a job can mean a side hustle, maybe you bake cookies and sell them, deliver pizza, freelance write, or whatever it can be your MOST VALUABLE way to getting a bitcoin. Suddenly that doordash/uber/pizza delivery gig seems like a million dollars in the future bank :).

Plan a set amount you can put in each month, from your job and pay yourself first and make it automated if possible. Get some of that monthly cash into bitcoin every week, every two weeks or monthly, but do it. Here is a way to free up more cash. Make a list of un-necessary to live bills and subscriptions e.g. do you really need that Hulu, Soundcloud or could you live without it for a future bitcoin? Do you get a tax return every year? If so you can adjust your tax withholding so you don't get one. That will free up some money that you can put into bitcoin every week, two weeks, month or whenever you get paid. As soon as you get paid in fiat we want to do the following immediately. Go buy bitcoin and put into the passive activities we will learn later on in the blueprint. For now park that cash in bitcoin. Fiat is losing value every year, bitcoin isn't.

Secret Genius Bonus Hack: In 10 years from now you could look like a genius for doing this one! Find a job that will pay you in bitcoin. One that comes to mind is Overstock. They will pay your salary in bitcoin. Or if you have an existing business or side hustle set up to have all your products and services with your clients as a preferred option to pay you in bitcoin. Talk about a salary bump! Picture for every $1 an hour it really being a future $10, $25, $100 dollars an hour. You could look like a real genius getting this going now.

SEMI-ACTIVE EARNING ACTIVITIES

We are repositioning ourselves in this step with Semi-Activities, so we do things we are already doing but repositioning so when you do them you earn bitcoin..

All bills, purchases, utilities, groceries repositioned to earn bitcoin.
Reposition all gifts to get bitcoin. Reposition all gas, electrical, gasoline, groceries all crypto back. Take advantage of all the spare things, and space you have as you'll see they are all assets. They just need to be activated so they generate bitcoin.

10 free mostly passive ways to make money with things you likely have already that can be exchanged for bitcoin.

1) Rent out your driveway on Neighbor.
2) Rent out your spare room, basement, garage to store people's stuff on Neighbor.
3) Rent out a room on Airbnb, Silvernest, Nesterly.
4) Pickup bikes and scooters and charge them on Bird.
5) Rent out your car, bike, scooter, curfboard, boat, atv, rv, swimming pool etc. on Turo, Getaround, Swimply, Outdoorsy, Boatsetter, Friendwitha.
6) Rent out your parking space on Curbify, Pavemint, Parqex
7) Put ads on your car on Carvertise, Wrapify and Promotocar.
8) Rent space to movies, photographers, events on Giggster, Splacer and Peerspace.
9) Rent camera equipment out on Sharegrid.
10) Rent out unused baby stuff on Babyquip.

Take all the proceeds and put it right into bitcoin.

Signup offers

This one is self-explanatory. Signup for an offer and get some free crypto. One thing to note is the qualifications to get the crypto. Some might say deposit $100 to get $15 in bitcoin, but you need to keep it there for 1 month or whatever is the qualification to receive the free crypto. Here are some good examples. ome of the places to do signup offers are; Coinbase, Tradestation, OKcoin, Nexo, and Celsius.

Contests/Drawings

Recent coinbase, ftx super bowl contest..didn't win boo hoo, but I tried. And if you don't enter you'll for sure never win. This will definitely not be your most consistent way to earn, but if you do then it will likely be a big payoff. Never enter drawings or contests that require you to send crypto to them. Only enter free drawings. There are some exceptions to this on legitimate sites and here is an example coinbase recently said if you buy $300 in Dogecoin then you can win $30,000. Well if you were going to buy some Doge anyway on a legitimate site like Coinbase, that might be a good contest to enter. However, many sites are set up as contests just to scam people, so know who you are dealing with and if in question don't do it. I've found typically it is the big exchanges like Coinbase, FTX, Kraken, Gemini are the ones that usually run the more legitimate and larger drawings. If you want to go small time, some youtubers regularly give out crypto to their subscribers for liking commenting etc.

Cryptoback

Occasional purchases like wedding, birthday gifts, holiday shopping, computers, laptops, earbuds, shoes, a washer and dryer, etc. All these should be done on an app or website calle Lolli. They have major brands you already likely use for major purchases. Brands like Bestbuy, Nike, Sam's Club and the like. Why not next time you need a pair of earbuds, order them on Lolli at Bestbuy and get crypto back for something you wanted to buy!

Need some dog or cat food, then order it on Lolli from Chewy and you'll be barking up some bitcoin :). Doing a renovation or that big landscaping project? Then head on over to Lolli and buy through Lowes, and you'll nail down some bitcoin! Need some sneakers, well sneak on over to lolli and buy from Nike, it's a slam dunk. Fluz is another up and coming place to do the same. Although Fluz rewards in fiat, it's simple to convert fiat to crypto and it has high % cash back.

Play2earn Games

Now you literally can get paid to sit around in your pajamas and play video games. Earn some crypto playing games. Some people are gaming their way to hundreds of dollars in crypto a day just gaming. There are all kinds of games from active games, to word games to just plain action figure games where you can earn crypto and NFTs. Here is a list of some popular ones; Townstar, Axieinfinity, The Animal Farm, Sandbox, and Splinterlands. Bitcoin Blast, blockchain game, and Bitcoin Pop some of the only ones on Google Play and App Store, and these are considered casual games, and they will be detailed out in the next section.

Play2Earn Game Tip: We would highly recommend playing games that have a solid market cap and established liquidity to trade the coins. Last thing you want to do is waste time earning some worthless coin that isn't tradable. So here is a way to get engaged in games that are reputable, popular and have the ability to exchange earnings out to bitcoin. The tip of the day is look at the games by market cap and invest your time in the ones that have an established base. This gives a list of the top ones sorted by market cap. This can be looked up on Coinmarketcap in the play to earn section of the site.

Casual Games

Casual games or Casual gaming has been one of the biggest trends in the gaming industry. It's a bunch of indie (independent) game creators developing typically addictive apps and simple web friendly games. Now casual gaming has entered the crypto space. Traditional games are controlled from a central location. In other words, a game's characters, skins, weaponry, and all coding cannot be reused in another game. Crypto gaming, on the other hand, many games allows everyone involved in the game to own a piece of it in the form of NFTs. Characters and other resources may function with other games if they are integrated with a the same blockchain or using a cross blockchain integrated game. Some are as simple as a little app you can get on the app store where you pop bubbles for prizes.

How Can You Make Money With Crypto Games In A Practical Way? Crypto casual gaming is a decentralized kind of gaming in which players can own unique in-game assets and sell them to anybody who is interested for real-world money. To put it another way, digital assets can be swapped for cryptocurrency, which can then be converted into real money. In a word, crypto games are created by storing information about unique assets that are totally held by players utilizing a whole or partial blockchain system. This is how consumers profit from cryptocurrency games. There are also a number of more simple games available on app platforms, where you can earn bitcoin. These Bitcoin game sites are entertaining ways to earn money on the side. The more games a person wins, the more Bitcoins he or she might earn. Some are dull, and some can be quite addicting.

Here are some of the available simple casual games you can find on Google play or the app store, and an overview of the objectives of the games.

1. CryptoPop A Bitcoin-based Candy Crush-style game. To play, tap on groupings of cryptocurrencies to pop them, such as Bitcoin, Ripple, Monero, and Ether. The more coins you can pop in a single tap, the more points you'll receive. Naturally, the more points you earn, the more money you will receive. The objective is to leave as few coins on the board as possible. Before you begin earning bitcoin, you must first input your Coinbase email address in the Wallet area in order for your points to be recorded. If you do not already have a Coinbase account, you can create one for free. CryptoPop currently supports Ethereum and Popcoin as digital tokens. However, the game will feature a few more cryptocurrencies in the future, including Bitcoin Cash, Litecoin, Binance Coin, and Dash. When you have enough points, head to the Wallet section and hit the claim button to receive your money.

2. CropBytes If you grew up playing Farmville or other online farming games, you'll undoubtedly enjoy CropBytes. You can play the role of a farmer, trader, or investor in this crypto game to expand your farming business and earn crypto on a daily basis. To make money from your farm, you'll need to first plant and harvest crops, then feed your livestock and collect items like milk, wool, and eggs. When you have a sufficient number of products, you can trade or sell them for cryptocurrency. The money you save can then be re-invested in your farm to make it even bigger and more successful. To get started, you'll need to create an account. CropBytes will give you free trial assets after you sign up, which you may use over the course of seven days. When your trial period ends,

you'll need to purchase in-game assets in order to continue building your business and earning cryptocurrency.

3. Bitcoin Pop It's simple to play; all you have to do is blast the unicorn's bubble into a cluster of bubbles of the same color. The more bubbles you bust in the fewest moves, the more Bling points you get and the more Bitcoin prizes you receive. However, keep in mind that in order to earn a big quantity of bitcoin, you'll need a lot of Bling points, which is normally the case with these crypto games. For the sake of comparison, 1,000,000 Bling points is equivalent to 0.0001 bitcoin. To begin playing, simply register for a free Bitcoin Pop account and log in. After that, simply input your Coinbase email address when you've earned enough bitcoin.

4. Bitcoin Bounce 15 Bitcoin Bounce is another Bitcoin game you may play to win cryptocurrency. The objective of this game is to bounce from one platform to the next as far as possible without falling. There are also bonuses and power-ups to help you survive longer and gain more points along the road. THNDR Tickets can be found as you travel deeper into the abyss. You must collect these tickets in order to enter the game's daily lotteries, where the real prize is found. Every day, the game distributes bitcoins to lottery winners. You have a larger chance of winning if you have more points and tickets, so playing frequently is beneficial. If you're chosen, Bitcoin Bounce will pay you via the Lightning Network, so make sure you have a Lightning Network-compatible external wallet. Wallet of Satoshi, ZEBEDEE, Breez, Bitcoin Lightning Wallet, and Blue Wallet are some of the wallets that are compatible with the Lightning Network.

Apps

There are several apps that you can download. Some even pay you in crypto to download and install them and give them a try. That's called paid per install. There are numerous and various ways to earn money on several different types of apps. Here are some of the apps to take a look at; Toluna, Unifimoney, Smilesbitcoin, Lympo, Lolli, Stormx, Featurepoints, Bitfortip, Blocklancer and Ethlance.

Faucets

A crypto faucet, sometimes known as a Bitcoin faucet, is a mobile app or website that rewards users with small amounts of BTC (satoshis or sats for short), or other cryptocurrencies in exchange for completing simple activities. You can get money by playing Bitcoin games, watching product videos, completing captchas, taking surveys, visiting links, and viewing advertisements, among other things. It's named a faucet because the rewards it disperses are similar to tiny drops of water dripping from a leaking faucet.

Bitcoin faucets won't make you rich by any means, but every sat can count, especially when those sats are becoming more valuable over time. While earning money with a free Bitcoin faucet may appear to be a piece of cake, depending on the faucet you're using, earning a substantial amount of cryptocurrency can take a while. The prizes you receive by performing various tasks will usually be sent immediately to an online wallet offered by the company. It's also worth noting that the lower the incentive, the easier the task.

Some faucets also have a minimum payout requirement, which means you can only withdraw your winnings once you've reached a particular amount. This could take a few hours, a day, a week, or even months to complete. Everything depends on the website's prices.

Now that you know what these faucets are and how they work, you're probably wondering why they were created in the first place. Here are some of the possible reasons why Bitcoin faucets exist today. Crypto faucets have been operating for over a decade from the time of writing this book, believe it or not. Senior Bitcoin engineer Gavin Andresen created crypto faucets to increase awareness about this new concept and sort of money a few years after Satoshi Nakamoto released BTC. Bitcoin faucets were created to distribute a prize of five Bitcoins for each operation completed—roughly $270,000 USD today!

Cryptocurrency exchanges were scarce in the beginning. Getting your hands on cryptocurrency in the early days of Bitcoin was difficult. There were few exchanges and venues where small amounts of BTC and other cryptocurrencies could be purchased quickly. To introduce it and stimulate people's interest without requiring them to buy, crypto faucets were used to distribute small amounts of Bitcoin for free.

Today, many BTC users supplement their income by setting up their own faucets. Owners frequently deposit a set amount of BTC into their crypto wallets, which are linked to their faucet website or app. Owners of faucets can earn passive income from advertisements as long as they produce more money than they are dispensing to consumers. Crypto faucets, in addition to serving as an early incentive system, distributor, and marketer of Bitcoin, have also worked to educate people about the benefits of BTC. These faucets can be a supplement to making passive money although admittedly it's slow and for some won't be worth the effort.

How Do Crypto Faucets Work? As previously said, Bitcoin faucets reward you with little amounts of BTC in exchange for doing small activities. You'll get paid in satoshis, which are the smallest unit of Bitcoin and named after its mystery founder. What is the size of it? One bitcoin can be divided into 100 million satoshis. The amount you'll receive is determined by the type of task you'll perform and the faucet provider you choose. It's just a case-by-case situation. I recommend having a separate wallet for faucets. Some exchanges can block receipt of bitcoin from these faucets. If you don't already have a a second faucet friendly Bitcoin wallets, you can sign up with many different wallet providers and get one for free right away.

Before you choose a Bitcoin faucet, consider the following factors; The next step is to find a Bitcoin faucet that best suits your requirements. But, before you make your decision, there are a few things you should keep in mind.
- Timers - Many cryptocurrency faucets include a refresh or loading period that might range from 15 minutes to many hours.
- Claim amount – This shows you how much and how often you can earn. This fluctuates depending on how simple or hard the work you'll be completing is.
- Minimum withdrawal - As previously said, some crypto faucets will demand you to reach a particular amount of satoshis before withdrawing your funds. Many faucets have a minimum withdrawal amount of

10,000 satoshis, which is about 0.0001 BTC or about $5 USD.

- Withdrawal method - You'll need a digital wallet to withdraw your funds.

Here are some popular crypto faucets—in no particular order—for you to check out.

Cointiply is one of the most profitable Bitcoin faucets and reward services around. By doing tasks, short offers, surveys, and viewing films, you can multiply your coins by up to 61x and earn BTC. This faucet also includes a loyalty incentive, reward points, quick payments, expert support, 25% referral claims, and more. Android users can download Cointiply's app.

Bitcoin Aliens By playing simple smartphone games, you can earn cryptocurrencies such as Bitcoin and Litecoin (LTC). It's been around since 2014 and has a 10,000 satoshi minimum withdrawal rate. Depending on how frequently you play games, you can earn up to 0.25 USD per day. Bitcoin Aliens is a game that can be played on Android and iOS smartphones.

Crypto Faucet 11 If you enjoy role-playing games (RPGs), Faucet Crypto is a good place to start. It has an RPG game style and other unique features, such as things that grant you various types of account perks. It also provides a direct withdrawal option, a 20%+ referral commission, and a lifetime level-up system that boosts your benefits. Faucet Crypto is also great for people who want to experiment with a number of different cryptocurrencies. BTC, LTC, Bitcoin Cash (BCH), Bitcoin SV (BSV), Dash (DASH), DigiByte (DGB), BitTorrent (BTT), Dogecoin (DOGE), Ethereum

Classic (ETC), Ethereum (ETH), Komodo (KMD), Tron (TRX), and other cryptocurrencies are all supported.

Fire Faucet-This faucet has a daily ranking system, where the top 20 users of the day can win enormous prizes; a level system, which allows you to raise your bonus as you level up; a referral system, and much more. Fire Faucet, like Faucet Crypto, accepts a broad variety of cryptos. BTC, ETH, DOGE, LTC, DASH, TRON, DGB, Tether (USDT), and Zcash are examples (ZEC).

Coinpayu- By just seeing numerous commercials and completing various offers, Coinpayu allows you to earn cryptos like as BTC, ETH, LTC, BCH, DOGE, DASH, and more. It also includes an affiliate program where you can make money by referring new people to the platform.

BTCClicks- This Bitcoin faucet is mostly focused on advertising, making it suitable for both earners and advertisers. It allows you to earn 0.00003 mBTC (0.0017 USD) for each ad and referral click you receive. Its affiliate program pays anywhere between 40% and 80% in commissions. BTCClicks also has a minimum payout of 0.10000 mBTC, which is approximately $5 USD.

If you wish to use BTCClicks to place an ad, the cost-per-click might start at 0.00006 mBTC or 0.0034 USD. While this faucet does not require user registration, it is vital to know that in order to begin earning, your audience must view your ad for the entire time of 10 to 200 seconds.

Satoshi Quiz Satoshi Quiz, This Bitcoin faucet rewards the first three persons who properly answer the question on the screen with modest amounts of BTC. Keep in mind that the question changes every 60 seconds, so provide the correct answer as soon as possible! At the time of writing, Satoshi Quiz offers a minimum withdrawal rate of 11,000 Satoshi

If you want to scale out your fauceting, then 99bitcoins keeps an up to date comprehensive list of faucets all in one convenient list.

Ad Watching
Imagine watching videos and getting paid in crypto. Well you can. Keep in mind payouts will be small, but every sat counts right? :) Some places to get paid for ad watching are the earn section on Coinmarketcap, Odyssey, Cryptosurf, Permission, and Playnano.

Browsing
Do you use the internet and search for things on a browser? (lol) Now switch browsers and get paid in crypto for doing it. If you want to switch your browser to start earning crypto that can be exchanged for bitcoin, then head on over to Brave.

Learning
Get educated on various crypto assets and you'll get paid on coinmarketcap and coinbase. Learn, take quizzes and earn. There is an earn and learn section on both Coinbase and Coinmarketcap.

Trading
I'll be honest, the vast majority of people shouldn't trade. Most are better off being long-term buy and hold or in this case buy and HODL investors. With that in mind, I've come up with 3 strategies over literally years of trading for all types, including the long-term buy and hodlr. I've looked at

backtests over several years to optimize what is the best indicator to use for buying and selling bitcoin. I've come up with one swing type strategy, and one long-term stack and hodl strategy I call my secret sauce. My secret sauce strategy is what I've termed my hodlstack strategy (combination of hodl and stack). Now as a word of caution, there are some things to avoid in trading. One is using huge leverage. Leverage is like salt, a little goes a long way. Too much though, and you ruin everything. Another thing to avoid is overtrading. Overtrading is trading too frequently. Brokerage firms and exchanges love when people scalp and daytrade, the vast majority pay fees for every trade and shortly blow their accounts. The secret of trading is to put probability and risk on your side.

I personally trade only on Kucoin. Why? For one reason only and that's they are the only ones I know of that offer a brilliant product that allows you to trade bitcoin with 3X leverage and will never liquidate your account. 3X is about the maximum leverage you would ever want to use. Anyone encouraging ridiculous 10 to 100X leverage has likely blown their account several times. Run for the hills if someone suggests trading with more than 3X leverage. You can 3X leverage long or short. The no liquidation is a great feature and it's why it's the only place I trade personally. Of course not financial advice and do your own research for your own situation.

Swing Method
Use a chart for this trading methods. I recommend using Tradingview. Set a 23 Moving average indicator and nothing else. Set the chart to daily. This was the optimal back tested strategy for swing trading bitcoin. When the price is above the 23 day moving average, then buy. When the price is

below the 23 day moving average, then sell. That's it and it's simple. It doesn't always work of course. You will have times where there are fakeouts. I usually like to wait for 3 daily candles to close before entering to minimize these fakeouts. Do not use the 23 day method with other cryptos. This is only for bitcoin and really that's the only coin with a volatility that's tradeable for most traders.

Trader Hack: For direction, look at the USDT and USDC dominance it correlates opposite with the price. This is a good indicator for finding direction of the market. When people start moving to stable coins price drops, and when people move back into bitcoin, market price increases. You can lookup stable coin dominance charts with the USDT.D and USDC.D symbols on Tradingview.

Long Swing Method
Set two indicators on Tradingview, on a weekly chart. A 200 week Moving Average, and a custom indicator called the pi cycle top indicator (which has a great history of calling tops for bitcoin). Buy when it hits around the 200 week moving average, and sell when it hits the picycle tops. Rinse and repeat. This is a very long-term swing trade and carry strategy that works great for stacking more sats. Remember you have to take taxes into consideration, which is out of scope for this discussion, so consult with a professional on your situation. With this strategy you'll be buying and selling maybe once or twice in a few years, so it's very long-term but very effective.

HodlStacks Method (the buy and hodl method to stack more sats without trading)

This method is simple. In bear markets bitcoin usually touch a 2 or 3 standard deviation level. What does that mean in plain english you ask? Well we will set a chart on Tradingview to a weekly timeframe. Then we will put two moving average indicators on the chart. Set a 200 Week Moving Average and a 300 week moving average. In the bear market (one occurs every 4 years in bitcoin) we will set an alert on Tradingview to send us a message when bitcoin crosses the 190 day moving average. When price crosses the 190 area we want to start buying because bitcoin is super cheap, and we keep buying more on the way down from 190 days until the area where price reaches the 200 day moving average, or if it keeps going lower (which is rare) we can keep buying all the way down to the 300 day moving average. This way you only buy when bitcoin is at it's cheapest price, and just hold long term. Save your dollars and buy more every time this happens and hold long term.

High Ticket Affiliate Earning
High Ticket Free To Join Referrals to Accelerate Your Sat Stacking. Let me ask you something. Would you rather sell something where you made $10 profit and had to sell thousands of those items, or rather spend time on one sale and get $1000 dollars or more on each sale? Often it's the same amount of effort for a $10 sale as a $1000 sale. You might have a bit more involvement but it can be highly worthwhile and lucrative to start some sort of promotion of high-ticket selling. I have just that in the next technique and this one hits close to home. It's actually my own wife's and family business. So with this one you are part of my online family, so I'm hooking you up!

Mexbuilders -High Ticket $2500/per sale
My wife comes from a town close to Puerto Vallarta. She comes from a family of Abagados (attorneys and investors). They've been building homes, and apartments since the

1970s. My wife decided to start a company to build homes in Mexico catering to Americans, Canadians and other foreigners that wanted to retire somewhere warm for less cost of living, and less building cost. I convinced her that she should also take payments not only in cash or crypto. It started with my wife and I building our own home, and then another, then some apartments using all the connections her family has and in 2018 Mexbuilders was born from that. She set it up so people can pay over time as it's almost impossible to get a loan on property or a standard mortgage without being a Mexican citizen. People can live in mexico much much cheaper and retire to a place that's beautiful, has fantastic food, and salt of the earth people. Millions of Canadian, Americans and other foreigners plan on or are retiring to Mexico. You can now help these people who want to retire on little more than social security but can't, have some options to do it now. You can even refer yourself, if you want to retire early and build a house on the cheap in Mexico.

Imagine how fast you can get to a bitcoin at $2500 for each referral that builds a property with Mexbuilders. Mexbuilders is my wife's and family's company.

I've worked it out with my wife that anyone that bought this book to is automatically approved for the Mexbuilders affiliate program. You can do this from anywhere in the world too. All you need is a phone and internet and use your creativity to find people who want to retire, but can't because of cost. Find people that want to build a home in Mexico to retire on their social security check and can make financed payments to Mexbuilders in Crypto or US dollars.

Mexbuilders is licensed and registered in the US and Mexico. Has years of construction experience. Full teams of

Abogados (Mexican Attorneys), Architects (Arcitectos), Accountants (Contadors), and Construction crews (Abaniles) to complete any project. Customers don't need to know any Spanish or even a thing about Mexico to retire there. Although I recommend a vacation there in Mexico prior to building a home.

Step 1: Signup at Mexbuilders affiliate program on the Mexibuilders dot com site forward slash /affiliates.
Step 2: Promote your link.
Step: 3 Get $2500 in crypto for each successful referral that builds with Mexbuilders. Just a few of these referrals will help you stack those sats fast!
Step: 4 You can also refer yourself or any family members to get commission too. Ever dream of retiring early to Mexico and living the crypto lifestyle? It's a very cheap cost of living and a wonderful life on the beach with great food, and great people. Then be your own referral and get $2500 and be an owner of the product.

Highticketaffiliateprograms dot com, is a site that has a very comprehensive list of high ticket items you can sell to stack sats quick. They have a thorough listing of over 1200 high ticket programs you can promote.

Airdrops
Airdrops can be a way to get free crypto and nfts. What Is a Cryptocurrency Airdrop? In the cryptocurrency world, an airdrop is a marketing campaign in which money or tokens are sent to wallet addresses in order to raise awareness of a new virtual currency. Tiny amounts of the new virtual currency are given to the wallets of active members of the blockchain community for free or in exchange for a small service, such as retweeting a company's message.

Crypto airdrops are a type of marketing technique used by crypto entrepreneurs. It entails distributing bitcoins or tokens to existing cryptocurrency traders' wallets for free or in exchange for a little promotional fee. The airdrop is intended to raise awareness and ownership of the cryptocurrency startup.

Airdrops are a type of promotional activity used by blockchain-based firms to assist fund the development of a virtual currency. Its objective is to raise knowledge about the cryptocurrency project and to increase trading volume when it launches an initial coin offering on an exchange (ICO). Airdrops are usually advertised on the company's website and on cryptocurrency forums. The coins or tokens are only given to people who typically but not always already have cryptocurrency wallets, like bitcoin or ethereum. To be eligible for the free gift, the recipient's wallet must contain a specified amount of the crypto currency in some of the cases. Alternatively, they may have to do a specific thing, like write a blog post about the currency, connect with a blockchain project member, or write a social media post about the currency.

Airdrops can be one the riskiest method of earning cryptocurrency. It's more than most investors believe is worthwhile. Airdrops are used by developers to gain support for new cryptocurrencies. In a nutshell, they provide the free coin in order to encourage adoption. You may monitor the airdrop project's development via the Internet. They are frequently pushed by users on the company's website, social networking platforms, and other cryptocurrency news sources. It is critical to exercise caution with every new cryptocurrency project. Hackers frequently use phony airdrops and ICOs (Initial Coin Offerings). So use caution and my advice to newcomers is to stick to the more well-known cryptocurrencies, such as Bitcoin and Ethereum.

Note: coins acquired via airdrops may be taxed as well in most countries, by most tax authorities. According to the IRS, you must report based on the distributed ledger's fair market value on the day of registration (in most cases when receiving airdrops from digital wallets). It's not tax advice so check with your tax professional for what is applicable for your specific case and location.

A genuine cryptocurrency airdrop will never solicit capital investment in the currency. Its sole purpose is promotional. On the other hand, other crypto frauds include the transmission of micro amounts of Bitcoin or other cryptocurrencies to unwary victims, a practice known as "dusting." Users should always be on the lookout for unsolicited deposits to their cryptocurrency wallets.

Let's take a look into the many types of airdrops that exists. Here are the most common types.

• Standard airdrop A standard cryptocurrency airdrop is a marketing approach in which a certain amount of native coin or token is distributed to existing wallets. Typically, it's to market the brand and urge more people to embrace the asset, which is frequently done during the asset's initial coin offering. In general, you only need to create an account with the new project and enter your wallet address during the distribution event.

• Bounty airdrops are similarly a form of marketing, but recipients must engage in some form of promotional action

in order to obtain the digital item. These activities could include:
• Sharing a tweet or other social media message about the blockchain project
• By subscribing to the project's email newsletter

• Participating in a forum to debate and contribute to the project. It takes a little more effort to obtain a free token via a bounty airdrop than it does with a conventional airdrop, but the activities are often not strenuous.

• Exclusive airdrops distributes crypto currencies to a select set of individuals who subscribe to an airdrop aggregator. These third-party websites provide information on prospective cryptocurrency projects and forthcoming airdrop events.

• Holder airdrops are distributed to those who hold a certain quantity of another cryptocurrency in their wallets. Typically, the crypto project takes a snapshot of cryptocurrency holdings on a specified date and then allows users to claim an airdrop depending on their ownership at the time of the snapshot. For instance, Stellar is a cryptocurrency project that began in 2014. In 2016, its founders announced a proposal to airdrop $19 billion worth of lumen (XLM), the project's native cryptocurrency, to existing bitcoin (BTC) holders as a gesture of gratitude to the Bitcoin network. To collect the XLM from the Stellar airdrop, you had to authenticate your BTC holdings.

How To Receive Crypto Airdrops? Crypto airdrops can be a great way to add to your crypto portfolio without having to buy any assets using fiat currency. Some ways to track down crypto airdrops are:

• Conducting routine online searches for crypto airdrop on Google, Twitter, Telegram, Reddit, Youtube and other social media platforms.
• By subscribing to airdrop aggregators and enrolling in their special airdrops.
• Joining new platforms and their Discord channels in order to take advantage of their standard airdrops
• Keeping an eye out for up-and-coming initiatives in order to prepare for bounty airdrops on Coinmarketcap.
Taking advantage of an approaching airdrop is largely a matter of being informed and seizing opportunities as they present themselves.

There are a good many sites that announce these whitelists of airdrops early. Most of these airdrops are from new coins and companies so basically startups. Occasionally you'll get some bounty rewards from more established projects. Typically there are a set of tasks to complete when participating in airdrops for example, like and follow on twitter, telegram, put out a retweet of their project, or other similar types of small tasks. Here are some sites to get early whitelist access announcements to airdrops; Coinmarketcap in the airdrop section, Airdrops, Airdropking, and Airdropalert. Also Reddit, Telegram, Discord, and Twitter are good places to also find these airdrops. These happen regularly, so you'll want to check in frequently.

Note: Never send an airdrop money or crypto, it's likely a scam called "dusting". Common airdrop scams will say something like we will 2 to 10X your airdrop if you deposit to this address. Never do this, it is almost certainly a scam.

Metaverse Real Estate

Think about buying a plot of land in downtown NYC or the coastline of California. Well you might have missed that boat, but the metaverse has built out a second opportunity for a land grab on virtual real estate. The metaverse has created virtual worlds where you can buy, flip, hold and sell virtual land in the metaverse. Here are some of the top metaverse plots that you can go after. There are also many other virtual worlds that are up and coming that you can get at cheaper prices. Facebook as you well known has rebranded to invest in the metaverse space as Meta so it's a growing trend, bound to get bigger.

So if your in the market for some land on the moon, somebody like has some to sell to you ;) Here are some Metaverse lands to visit; Sandbox, Somniumspace, Cryptovoxels, Metahero, Bitcountry and Upland.

DAOs
Ever wanted to invest like an angel investor in new startups, but were priced out of that Wall street closed investment? Well now you can invest like the rich do in the new startups. The new startup organizations are called DAOs (Decentralized Autonomous Organizations). They are companies owned by those who participate in the DAO. It's an organization without an owner, and you can go get a job working for a DAO that has no boss, no owner, and take part in decentralized voting ownership on how the organization operates. You earn money by helping move the dao forward by hodling their crypto coin, and/or contributing to the organization. Lists of DOAS are kept on Daolist and on

DeepDao. You could work for starting 5 minutes from now with no interview required as most are open and permissionless.

Want to go full time working for DAOs? Here's a list of some; Metacartel, MetaClan, Poap, Daohaus, Fire Eyes, Rocketpool, Metafactory, Seedclub, Friends with benefits, Audius, Forefront, Gitcoin, Universe, Fingerprints, SperRare, Pleasr.

OK now for the fun part, and now that you have the Active and Semi-Active Activities going in full-swing at this point. It's time to take the money you made and are making and get it flowing into the Passive side (so that money grows and works for you even while you sleep or go to the beach all day).

PASSIVE EARNING ACTIVITIES

Now for my favorite way to earn, and that's doing literally nothing. What we want to do is take the crypto earnings from ALL of the Active Activities and Semi-Active activities and get it earning money on top of money or in this case bitcoin on top of bitcoin. So all earnings flow into the passive earning area. Whenever we have any income from a job, active side hustling, or anything else, we want to park it here in the Passive area as soon as possible, so that we are always compounding. This allows for growing that bitcoin even when we're sleeping or building sandcastles on a beach.

Staking:
Think of mining and staking like creating bitcoin. Similar to how to create a real coin, a Mint stamps out some metal to create and forge new coins into existence and circulation. Mining uses electricity to power computers and devices to solve cryptographic puzzles to create bitcoin. Staking is sort of a different approach to mining. Instead of using electricity and power to create crypto currency by solving algorithmic puzzles like bitcoin does. Some coins create coins by staking, which is a process to prove you have ownership of a coin, and thus secure and contribute to the network. Now if you didn't understand a word of that, don't worry you don't need to. Today staking is as easy as depositing money into checking and withdrawing it when you want to. Staking is buying a crypto and then parking it on an exchange or in a wallet, and it earns interest, just like your checking account except much, much better :).

How Crypto Staking Works. Crypto staking is how new transactions are added to the blockchain in cryptocurrencies that follow the proof-of-stake paradigm. Participants first make a pledge to the bitcoin protocol using their currencies. The protocol selects validators from among these individuals to confirm transaction blocks.

You're more likely to get chosen as a validator if you pledge more money. New bitcoin coins are produced and paid as staking rewards to the block's validator every time a block is added to the network. The rewards are almost always the same type of coin that the players are staking. Some blockchains, on the other hand, use a different form of cryptocurrency as a reward. You must own a cryptocurrency that uses the proof-of-stake model in order to crypto stake.

Many popular bitcoin exchanges allow you to do so. When you stake your coins, they remain in your possession. You're effectively putting those staked coins to work, and you may unstake them at any time if you want to exchange them. The unstaking procedure may take some time; some cryptocurrencies require you to stake coins for a set period of time. With all sorts of cryptocurrencies, crypto staking is not an option. Only cryptocurrencies that use the proof-of-stake model are supported. To add blocks to their blockchains, many cryptos use the proof-of-work concept.

The issue with proof of work is that it necessitates a lot of computational power. As a result, cryptocurrencies that use proof of work consume a lot of energy. Bitcoin (BTC) has been chastised because of environmental issues. Proof of stake, on the other hand, necessitates far less effort. This also makes it a more scalable solution that can manage higher transaction volumes.

How To Do Crypto Staking Crypto Staking may seem a little confusing the first time around, but it's a simple process once you get the hang of it. Here's how to stake crypto step by step:

1. Purchase a proof-of-stake coin. Staking is not available in all cryptocurrencies, as previously stated. You'll need a cryptocurrency that uses proof of stake to validate transactions. Here are a couple of the biggest cryptocurrencies you may invest in, along with some information about each: Ethereum (ETH) was the first cryptocurrency to have a programmable blockchain on which developers could build apps. Ethereum began as a proof-of-work system, but it is now shifting to a proof-of-stake paradigm. Cardano (ADA) is a cryptocurrency that is beneficial to the environment. It was produced using evidence-based procedures and was based on peer-reviewed research.

2. Put your cryptocurrency in a blockchain wallet. Your cryptocurrency will be available in the exchange where you purchased it when you purchase it. With certain coins, several exchanges have their own staking mechanisms. If that's the case, you can simply stake crypto on the exchange itself.

3. Participate in a staking pool. Staking can vary depending on the cryptocurrency, but the majority use staking pools. To increase their chances of receiving staking rewards. Pools are not a necessity, and now exchanges do all the backend work, and have made staking as easy as it is to deposit into a checking account as an analogy.

Staking is powerful. Here is a scenario..let's say you could get enough earnings to buy 300 Polkadot, symbol DOT and stake it on Voyager or Kraken. Earning 12% staking rewards, by 2030 given some projections of DOT hitting $100 you could literally have almost 1 million dollars in that scenario. That's likely enough future dollars to have the equivalent of a bitcoin. I have 2 rules when staking. First, I look for high quality alt coins like ETH, Solana, Polkadot that are top 20 coins on coinmarketcap. I also look for staking rates that return 5% or more. The majority of my staking is with Polkadot and Solana. It's growing in value and earning interest even when I'm sleeping or building sandcastles on a beach. Some legitimate places to stake crypto are Kraken, Binance, Binance US, Kucoin, and Voyager.

Welcome to the Wild West of Crypto:
In this next section we are getting into what I lovingly call the Wild West of Crypto. In these next sections we are going to discuss DEFI(Decentralized Finance) and CEFI (Centralized Finance) and some potential risks and rewards in both of them. This CEFI/DEFI/Yield Farming is the dark underbelly, the dark alley way of crypto so you may decide to mosey on by this section if you don't have an appetite for ridiculous APY and high risk. It's all Lions, Tigers, Bears, Sharks and Impermanent loss Oh my! If you are up for it let's crack on.

Hacks, and scams with smart contracts, impermanent loss, opportunity loss reign here, so here is a rule to live by if participating in the underworld of DEFI. IF you see an APY above 20% it is to be considered extremely high risk. Never put anything in you are not willing to ever see again. Now with that said, IF you decide to do this, ONLY use a very small amount of your Passive money. As you go farther out on the risk curve doing DAO earning, Copy Farming and Yield Farming, it's a very risky world. So as you go out on the risk curve allocate less of your passive capital to those things (if you decided to do them at all). Next time you see some shady offer that's giving out 150,000% APY, and a couple of watches out of his trench coat ;)...don't say you weren't warned. Most people get lured in at one time or another, and hopefully not too many times before they learn the lesson. Like I always say, greed can be an excellent teacher to the right student.

CEFI (Centralized Finance)- Lending

Think of centralized finance similar to a bank. Banks are owned by people and or companies, and that is exactly what centralized finance is. You are lending out your crypto and earning an APY and they are lending it out on their platform. The benefit of CEFI is you can actually pick up the phone or chat with the company for support, whereas in DEFI it's typical that there is none as everything is run by contract and code. Note, I like to spread my risk out to multiple CEFIs, and DEFIs. Of course not financial advice and you should always do your own research. Here are some popular ones; Blockfi, Ledn, Voyager, Nexo, Circle, and Vauld. A word of warning when you put your money in a CEFI, you lose an element of control. If a run happens to try to withdrawl, these sites can be risky because there is no regulation to say how much capital they have to keep in their reserves. Some of these CEFIs have made the news by pausing withdrawls when these run on the exchange happens. I typically prefer storing my crypto on a hardware wallet like ledger and delegate staking on something like DOT or Polygon. That way I'm in control of my crypto at all times, and make comparable returns.

DEFI (Decentralized Finance)-Liquidity Providers

The banking system is slowly being replaced by crypto banks that lend, and provide earnings, but they are not owned by a company or person. This is DEFI. Imagine becoming bankless, or better yet becoming your own bank with DEFI, and instead of paying the man, you earn the interest for once. You are providing the assets (crypto), known as being a liquidity provider. Those assets have a demand from people needing liquidity and loans, and this produces an interest yield to you. Welcome to defi where you can borrow without credit scores, and lend and collect interest. Here are some popular DEFI exchanges that can keep your money growing, even while you are building sandcastles on the beach; Ribbon Finance, Stacks Co, Thorchain, Cakedefi, Aave, Compound, Meld, Beefy Finance, Orca, Pancake Swap, Sundae Swap, DefiLama.

Arbitrage

Arbitrage in it's most basic form is simply buying low and selling high. For crypto it's a process of buying an asset at a higher or lower rate on one exchange and offsetting it with the price of the same asset at a higher or lower different rate on another exchange. Bots have been programmed to constantly scan crypto exchanges for price differences and then collect the difference between the prices as profit. Some popular arbitrage bots and platforms are; Whitewhale Money, and Pionex.

Copy and Yield Farming

Crypto yield farming, also known as yield farming, is the practice of lending bitcoins to exchanges in exchange for hefty fees. Liquidity mining in a liquidity pool allows you to put your digital assets to work. Typically, this dividend will be paid out in cryptocurrency. It does, however, necessitate both a liquidity pool and a liquidity provider. Due to the multiple benefits it brings to the exchange, it is a part of decentralized finance (DeFi), and this is the primary reason for its widespread appeal. You can farm with multiple crypto assets and obtain crop yield through DeFi lending within the DeFi ecosystem by employing the borrow and lend mechanics. Many investors might stake stablecoins like USDT or USDC when yield farming was originally offered. How To use Yield Farming To Earn Cryptocurrency Yield farming relies on an order-matching mechanism known as the automated market maker (AMM) model to function. The automated market maker (AMM) model is used by the majority of decentralized exchanges. Rather than specifying the current market price of an asset, smart contracts are used to create liquidity pools. The pools can then use the preset algorithms to execute the trades.

Liquidity providers must deposit monies in the liquidity pools for this operation to work. DeFi users can borrow from, lend to, and trade through the pools, which offer the finance infrastructure. Users must pay trading fees, which are then divided among liquidity pools based on the amount of liquidity they can supply. Polygon, for example, has yield farming that is particular to Polygon, BSC yield farming (Binance Smart Chain), and so on. Yield farming DeFi methods are the umbrella term for all of them. Because there are market makers on Binance, yield farming is considered one of the finest choices. Yearn. Finance is also a cryptocurrency yield farming platform.

It's worth noting that there are cross-chain platforms that support many coins and DeFi initiatives. Furthermore, there are various DeFi tokens that enable you to act as a liquidity provider on the Ethereum network. Stacking is similar to mining, but the main distinction is that you keep your coins in a crypto stack rather than mining them. A transaction can also be validated through stacking. In mining, the first person to solve the puzzle is the one who gets to record the transaction. These are chosen at random in stacking; whoever has the most stacks has a better probability of documenting the transaction, and the individual is rewarded for doing so. You must have a verified account and a crypto wallet in order to stake your cryptocurrency. This is a more environmentally friendly approach to earn essentially free bitcoin.

Defi yield farming can be difficult to navigate, Ethereums high fees also complicate matters, and with impermanent loss, multiple wallets and tokens. Because of this the concept of Copy farming is emerging as a solution that might be better suitable for the majority of people when entering the world of yield farming. Copy farming allows someone else to do the yield farming for you, or even a programmed machine learning optimized program to do all the yield farming for you so that you are freed from the complexity of doing it yourself. Copy farming basically allows someone else or even algorithmic code to farm for you, and then they take a commission which is typically 5-10%. Here are some popular ones; Donkey Finance, Orca, APY Finance, and the Animal Farm.

Bots
Bots are automation used for trading crypto or using an arbitrage strategy. The pros of a bot is they eliminate the emotional reaction typical in trading, and they really stick to the plan. The bad news about bots is they stick to the plan,

and don't intervene in scenarios where you might want to get out of a trade and cut a loss. Bots can and do work, but the vast majority don't work, so use money you can afford to lose if you go there. Personally the only ones I use are listed here and I use very small amounts of capital, as it's out on the far end of the risk curve for me. Of course and with everything in this but, it's not financial or any other advice. Always do your own research for your own situation. Here are some popular crypto bots to look into; Pionex, Kucoin in the bots section, Cryptohopper, and Haasonline.

Mining

By mining bitcoin, you can earn it without having to pay for it. Bitcoin miners are rewarded with bitcoin for completing "blocks" of validated transactions and adding them to the blockchain. Mining rewards are distributed to the miner who solves a complicated hashing puzzle first, and the probability that a participant will solve the riddle is proportional to his or her share of the network's total mining power. To set up a mining setup, you'll need either a graphics processing unit (GPU) or an application specific integrated circuit (ASIC).

Mining is a critical component of the Proof of Work (PoW) consensus mechanism and is one of the most established methods of generating revenue with cryptocurrencies. This is the process through which transactions are validated and a PoW network is secured. Miners are rewarded with new coins in the form of block rewards for performing these functions. Mining was possible on a desktop computer in the early days of Bitcoin, but specialist mining hardware is now a necessity. Don't bother mining with these places that say mine on a phone or your desktop. These are not recommended ways to earn.

.

There are primarily two types of mining; cloud and hardware mining. With hardware mining, you are more in control of the outcome. However, with hardware mining there is an investment in upfront money, time, electricity cost for a payback of the mining equipment. Also, mining offerings on phones and other devices are typically and honestly just not that profitable. You could be mining for years and earn pennies on those apps, so I don't recommend any that I have seen or tried (change my mind). I don't recommend mining bitcoin for small miners, because the difficulty and profitability just does not work unless you've already paid off some miners and already been doing it a long time. Unless you are capable of running a very large operation with cheap power and hundreds of miners I would not bother at a small scale at this stage of the game. Not to fret, there are a couple of other options I see as still potentially profitable in mining. Two very legitimate alt coins that could be profitable for a small miner are Helium and Ravencoin. Both can be exchanged for bitcoin. Helium uses a small radio device and your wifi connection to hook into a network, and the mining rigs are cheap enough to be paid off quickly, and use little energy. Ravencoin is also early enough and cheap enough to mine profitably. To mine on Helium you'll need a miner and those can be found at Bobcatminer.

Cloud mining is mining on a site, where they run the hardware and electricity for you and give you a share of the mined crypto. No hardware required with Cloud mining. The longest running and most credible are Genesis Mining, and Viabtc.

Note: Cloud mining is riddled with scams so use caution. The two cloud mining companies that have been around for years are viabtc and genesis mining. They often sell out of contracts but check back often as they do become available from time-to-time.

Bonus Mining Hack: There is a guy that lives by me that's made quite a business of helium mining. He contacts businesses in person and proposes to let him put in a helium miner on their wifi. The business owner typically likes the idea because it helps them offset their lease rent, and he likes it because he splits the monthly profit on the mining with not paying for the bandwidth and geolocation. Talk about scaling the operation.

Dividends

Dividends are another way to earn crypto with. If you've ever invested in stocks or bonds, you're definitely familiar with the concept of dividends. Dividends are small financial payments paid to shareholders, to put it another way. If a firm earns profits for a quarter (or a year, depending on whether it is a sole proprietorship), the profits are split and returned to the company's ownership (shareholders). While a tidal wave of dividends may not hit your crypto account unless you have a substantial amount, it can be a method to make money with the crypto you already hold. You must, however, conduct a study to determine which cryptocurrencies offer dividends and whether the payouts are worthwhile. VeChain, NEO, Reddcoin, NAVCoin, Decred, and their annual dividends are examples of cryptocurrencies that pay dividends in multiple coins (or tokens). As a result, unlike stock dividends, cryptocurrency dividends distribute more tokens rather than cash. VeChain's Vthor dividends has been very consistent. VET or VeChain can be purchased on many exchanges including Binance or Binance US for US residents.

Masternodes

Maternodes are running the computers that help validate and secure the network for crypto. Masternodes typically require not only an investment in virtual computers, or physical computers, but also an investment in owning a large portion of the crypto coin you run a node for. However, there are platforms that do all of this for you and you split the profit from having them run the nodes for you. Some places that will run nodes with you and profit share are; Yieldnodes, Thor Financial, Strongblock, Flux, and Vapornodes.

Wrapping it all up

Thanks for buying us a coffee, and thanks for joining us in our cause to help as many people as possible earn a bitcoin! Most importantly follow us on Youtube, Twitter, Tiktok or Telegram to get updates, reviews techniques and all things to get you earning one bitcoin!
We are so grateful for each one of our subscribers and want nothing more than to see all of you get a bitcoin so join us on the journey!

Some Parting Inspirational Scenarios Just for Fun and Profit?

Motivational Scenario

If you can get 300 Polkadot staking at 12% you'll be well on your way to getting a bitcoin. Here is what that looks like over 10 years.

Staking 300 DOT: 300 for 10 years could yield about 900k. Assumptions are that Polkadot grows to $1000 a coin within 10 years, which given growth rates, demands and projections is not too out of the realm of possibility.

QUICKSTART GUIDE - 1 PAGE OF ACTION

Active

Step 1: Earn money with Job, Side Hustles, Microtasks, flip stuff, hold a garage sale --put it regularly into Passive Crypto earning (see Passive)

Step 2: Get a venmo or crypto dot com card and reposition ALL your purchases (utilities, groceries, bills, even house payment on the card and pay the full balance off every 21 days or before you are charged interest)- set venmo to get crypto back in the credit card settings of venmo.

Semi-Active

Step 3: Do some semi-active activities on a regular basis --put it regularly into Passive crypto earning (see Passive). Some semi-active activities are; Play2earngames, browsing, ad watching, earning apps, faucets, trading, airdrops, high ticket affiliate programs.

Passive

Step 4:

Take earnings from Active and Semi-Active activities and regularly earnings them into Passive earning..so your crypto keeps working, compounding and growing even while you sleep or build sandcastles on a beach. Some Passive options are; Staking, Defi, CEFI, Mining, Arbitrage

Bonus: If I had to pick only 3 things to do. I would do the following;
1) Reposition: Get a venmo card and get Ethereum back by setting up all my bills, groceries, dining, gas, utilities, internet and all other bills repositioned on venmo for crypto back.
2) Earn: Use my job and a high ticket program like Mexbuilders to make $2500 a sale.
3) Stake: Use my money in #2 to put it into Polkadot on Kraken staking and growing it at 12% interest. If you get 300 Polkadot and earn 12% in 8-10 years if Polkadot reaches $1000, then you would have around a million dollars or likely the equivalent price of one bitcoin.

Thanks for Joining Me On the Journey

Hopefully you enjoyed this book, and if you put even a portion of the steps into practice, it will more than pay for the book's cost for an almost certainty. If you enjoy the journey and the information I provided, please consider giving us an honest review. Thank you so much for taking the journey with me to Getting Your Bitcoin!

Disclaimers:

Financial matters are highly individualistic. Risk tolerance is just one factor to consider before making any investments or financial decisions. For these, and other, reasons, you should look to the guidance of a trained professional, not a website or a guide. All examples and offers are for educational and example purposes only.

Any information found on or in this guide or associated sites are not financial, legal, tax or any other advice. Research on your own or with a professional before making any decisions.